The Power of Physics

David Ash

"If we do discover a complete theory, it should in time be understandable in broad principle by everyone, not just a few scientists. Then we should all, philosophers, scientists, and just ordinary people be able to take part in the discussion of the question of why it is that we and the universe exist. If we find the answer to that, it would be the ultimate triumph of human reason - for then we would know the mind of God."

Stephen Hawking

PujaPower Publications
Parkhill, Littlehempston
Devon TQ9 6LY

0-9550857-0-5

I thank Liesbeth, my wife, for her love and support, my friend Philip for pressing me to write 'The Power of Physics' and Anna, my first wife, for her care and delightful line drawings!

This work is dedicated to the memory of my father Michael Ash who introduced me to nuclear physics before I could ride a bike.

David Ash email: ddashlash@aol.com

Erwin Schrödinger, *"If you cannot - in the long run - tell everyone what you have been doing, your doing has been worthless."*

Gary Zukav, *"The fact is that physics is not mathematics...Stripped of mathematics, physics becomes pure enchantment."*

Eric Laithwaite, *"I believe that matter itself is just spin."*

Welcome to the New Physics of Consciousness

Lord Rutherford of Nelson – where my beloved wife Liesbeth comes from – said, "These Fundamental things have got to be simple." You will love the innate simplicity of my account for the atomic nucleus and the proton – discovered by Rutherford - and the other particles of nature, as well as the forces that hold them together. Come on a journey with me and, through the new physics of consciousness, discover the Universe for yourself. Understand the world in which we live and other worlds beyond. Appreciate yourself as a multi-dimensional being in a multi-dimensional Universe. Know yourself. Realise your true essence. Discover your divine potential.

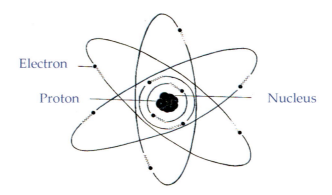

The Atom

Your body and the chair you are sitting on can be broken down into molecules. The molecules are then made of atoms - carbon, iron, hydrogen etc. These in turn are formed of sub-atomic particles - electrons, protons and neutrons. The question is, what are these sub-atomic particles?

Mass is Energy

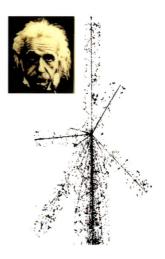

In his equation $E=MC2$ Albert Einstein showed that mass is equivalent to energy. High-energy experiments have proved that sub-atomic particles of matter can be formed out of energy. But how is a massive particle formed out of pure energy? The answer to this conundrum is found not in modern physics but in ancient yogic philosophy. Thousands of years ago yogis in India taught that energy - prana - exists in matter - akasa – in the form of vortices - vritta

Two Forms of Energy

Imagine energy as a line of movement. Think of the line undulating in waves to form a photon of light…

The photon as two wave trains of energy

…or spinning to form a sub-atomic particle of matter.

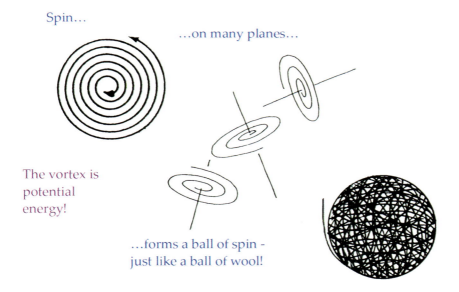

Spin…

…on many planes…

The vortex is potential energy!

…forms a ball of spin - just like a ball of wool!

The Spherical Vortex

If the smallest particles of matter were vortices of energy, what form would they take? As the line of movement of light is free to spin on infinite planes the vortices - like balls of wool - would be spheres. With the axes of spin constantly changing they would have no poles. You could visualise the particles as spherical vortices of light.

The Vortex explains Inertia

The vortex explains inertia very easily. Richard Feynman, one of the greatest Physicists of the 20th century said, "The law of inertia has no known origin."

We know spin creates inertia. The spin of a gyroscope sets up inertia. A spinning pebble skips across a pond because its spin creates inertia in the plane of spin. Spin on infinite planes would set up resistance to movement in all directions.

The Infinite Extension of Vortex Energy

Why is matter 3D and not 2D or 6D? Three-dimensional extension is a characteristic of the vortex. The 3D extension of the vortex would be conferred on all matter formed of vortex energy.

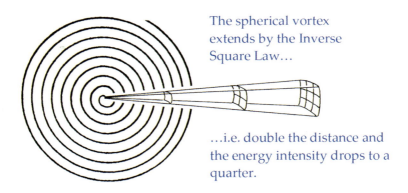

The spherical vortex extends by the Inverse Square Law…

…i.e. double the distance and the energy intensity drops to a quarter.

If energy is neither created nor destroyed the vortex of energy would extend into infinity. Look at the power of this idea!

Force Fields

Overlapping and interacting vortices of energy would account for the distant forces of matter such as electric charge, gravity and magnetism.

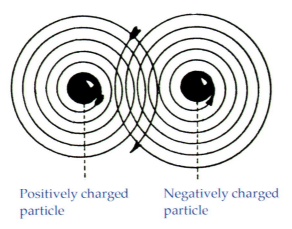

Positively charged particle

Negatively charged particle

For detail on electro-magnetism visit the Quantum Vortex on pujapower.com

Magnetism is caused by...

Electric charge is caused by interactions of overlapping concentric spheres of vortex energy that are either expanding or contracting - Yin or Yang.

...a rotating vortex

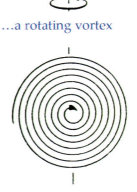

The infinite extension of the vortex accounts for the infinite extension of electric charge, magnetism and gravity. This implies that everything everywhere is connected to everything else!

Space

If matter were the vortex energy we perceive and space were the vortex energy beyond our perception, the infinite extension of the vortex would account for the infinite extension of space in three dimensions. This principle explains relativity. Einstein summarised his theory saying, "Move matter and you also move space-time." As you move, your 'space bubble' would move with you!

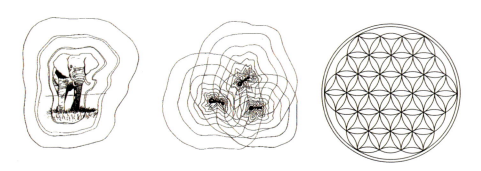

Every body creates a 3D bubble of space that extends into infinity.

Your Infinite Space

When you set up vibrations in your extension of space by thought and intent, your frequencies extend throughout the Universe touching everyone. They have a subtle influence on everything. This is how the collective thoughts of humanity affect the entire Universe. For example, war zones respond to our frequencies, influenced by whether we chose to transmit peace or anger.

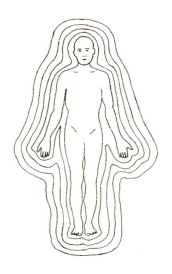

Materialism is Unscientific

In ancient Greece, Democritus, the father of the atom, invented materialism. Democritus taught the world is formed of indestructible atoms that move in the void of space. Einstein showed this is not so. Mass and space depend upon movement; they are relative to the velocity of light. Einstein's insight is readily explained by the vortex.

The Vortex explains away Materialism

Mass, inertia, forces and space are accounted for by energy in the spherical vortex. The Universe is formed not so much of particles that move as particles of movement. As a particle of movement the vortex explains away the properties of 'material substance'. The new physics of consciousness is the antithesis of materialism. It reveals scientific materialism as the 'illusion of maya'.

Curved Space

The space around the sun is curved as it is an extension of the sun. Starlight follows the curved space round the sun like a car on a roundabout. If you find this hard to grasp go check out the alternative, Albert Einstein's General Theory of Relativity! Einstein's theory is arbitrary complex and it doesn't fit with Quantum Theory. The vortex idea is axiomatic, it is simple, and it leads to a totally new, playful approach to Quantum Theory.

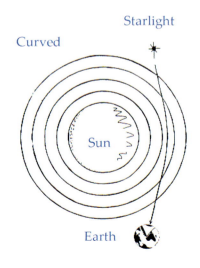

Starlight

Curved

Sun

Earth

Quantum Sex

Wave energy is like sperm. Treat light as masculine. The spherical vortex of energy is like the ovum. Consider matter feminine. The word 'matter' is derived from the Latin for 'mother'.

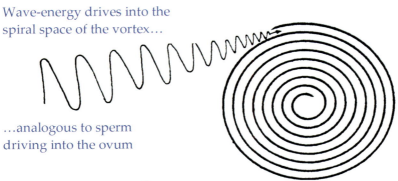

Wave-energy drives into the spiral space of the vortex...

...analogous to sperm driving into the ovum

Quantum Marriage

Unlike young men and old fools the wave train cannot whip in, whip out and whip away as the flow of energy is uni-directional. This is time – past to present to future. There is no reverse gear; no way out! In the resultant marriage of wave and vortex, kinetic energy is channelled into mass, work and atomic structure. The separation of wave and vortex leads to disorder i.e. entropy.

Wave-particle Duality

In the wave-vortex couple, a wave energy 'tail' propels the electron in wave motion - much as a tadpole is propelled by its tail. Wave-vortex coupling accounts for the wave-particle duality of electrons.

Quantum Hanky Panky

Lady electron has an energetic lover. She takes a quantum leap into the excited state. But the triangle is unstable. The 'old flame' leaves as a photon of light and she is 'grounded'. Soon she is off again but not just any old bloke will do for lady electron. He has to have what it takes to excite her to the next quantum level!

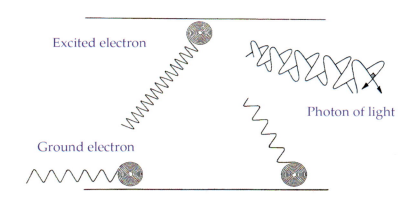

The Quantum Leap

Electrons in a flame leap into the excited state. Dropping back to the ground state they release energy as photons of light. Their colour represents a frequency - the energy difference between the two states. Only inside the flame do temperatures allow the 'quantum' packets sufficient energy to make the quantum leap.

I can quantum leap

The Mechanics of Quantum Sex

The quantum is gay; a two wave couple. Lady electron only gets the hots for wave singletons. She and hubby are swingers. The more single guys they take on the more excited they become!

Wave train & vortex

Gay quantum wave couple = h

Wave singleton = $^1/2$ h

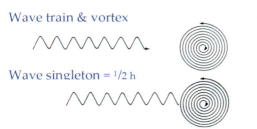

The quantum consists of two energy fields. The fundamental packet of energy is a single line of the movement of light. This is half a quantum. Because Planck's constant 'h' delineates the quantum, events in quantum mechanics taking on single wave packets of energy would occur as multiples of half h.

"I think I can safely say that nobody understands quantum mechanics."
Richard P. Feynman

Chemistry

If the chemistry is right and lady electron really gets the hots she will leave her atomic home with her lover and settle in another. The 'electric' pull on the old home is an 'ionic bond'. Nipping back and forth between the old atomic home and the new one is called the 'covalent bond'. This happens all the time in 'quantum love affairs'.

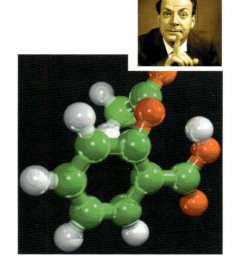

Captured Mass

Proton vortices are 1,836 times as massive as electrons and have sufficient inertia for wave trains of energy to drive right into them until their inner spiral space is saturated with energy as 'captured mass'. Proton vortices capture energy much as black holes capture light.

Nuclear Energy

When protons collide in nuclear fusion, or converge after nuclear fission, captured mass is displaced as nuclear energy. In fusion this is about 5%.

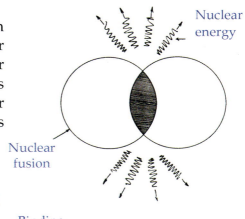

Nuclear energy

Nuclear fusion

Nuclear Force

The 95% of captured energy left behind swirls between the protons binding them together. As protons converge after fission, though more captured mass is lost, the shorter path of the captured energy tightens the binding. This account for nuclear energy shows why nuclear binding increases as mass is lost.

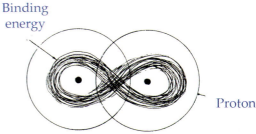

Binding energy

Proton

"Nuclear energy, we have the formulas for that but not the fundamental understanding. We don't know what it is!"

Richard Feynman

13

Anti-matter

When batter passes through a doughnut machine it takes on the form of a doughnut. When wave energy drives through a vortex it can take on the form of the vortex. This transformation of energy into mass is illustrated by the production of anti-matter. When a gamma ray photon drives through the vortices in the nucleus of a heavy atom one wave train spins into an electron. The other is transformed into an anti-matter 'positron'. When the positron meets an electron the vortices annihilate reverting to gamma rays

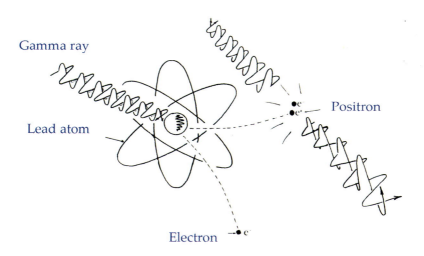

Gamma ray

Positron

Lead atom

Electron

The Mirror-symmetrical Universe

Everything in the world of matter could be replicated in a mirror-symmetrical world of anti-matter. Passage into the looking glass world of anti-matter would come by shrinking into the smallest space or expanding into largest space much like Alice when she ate the 'eat-me' biscuit or drank the 'drink-me' drink.

The Twins

The electron 'funnel' of vortex energy in the world of matter spins through a 'tunnel' of zero-space out into a positron vortex in the world of anti-matter. A common zero-space centre connects these twins in the 'Alician' dimension of bigness and smallness.

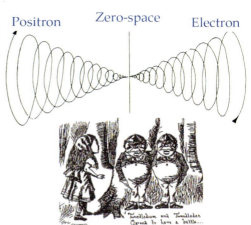

Infinity

Infinity is the circulation of vortex energy as in the water cycle. The centres of vortex particles are as raindrops. The single largest sphere of space is like the ocean. The Universe, in two halves - matter and anti-matter – connects through the centre of all the sub-atomic vortices, and also through the single largest sphere of space.

Gravity

Gravity is the force of attraction between matter and anti-matter through smallest space. We experience this as the 'centralising' pull between the matter in our bodies and anti-matter through the centre of the Earth. The pull between matter and anti-matter through the largest sphere of space causes the Universe to expand.

15

The Unified Field

The idea that particles of matter are vortices of energy fits with so many discoveries, observations and experiments in physics, the vortex could be the unified field Albert Einstein was searching for. Unifying relativity and quantum theories, the vortex points to new directions in science and upholds ancient knowledge.

"A theory is a good theory if it satisfies two requirements: It must accurately describe a large class of observations on the basis of a model that contains only a few arbitrary elements, and it must make definite predictions about the results of future observations."

Stephen Hawking

Understanding Energy

Mass is energy. Energy is movement but nothing is moving. Energy is more an act of consciousness than any thing. Particles of energy are more thoughts than things. If particles were thoughts then the Universe would be a mind! Could it be the Universe is conscious? Is the Universe the mind of God?

"It is important to understand that in physics today we have no idea what energy is!"   R. P. Feynman

Universal Consciousness

All protons appear to have identical characteristics, which suggests they could originate from one indivisible consciousness. If consciousness were indivisible we could be the one being in many bodies. Maybe this is why we are called 'Hu' - meaning 'God' - man beings? Perhaps each and every person who has ever lived is God incarnate?

My Prediction

The Universe is like the atom.

"As above so below,
As below so above."   Hermes

The Mandelbrot Set

The Mandelbrot set is a fractal. Each minute projection is expanded from the original pattern and is an infinite repeat of the original pattern.

The Fractal Universe

Maybe we live in a fractal Universe where patterns repeat at every level. The patterns we observe in the atom could repeat throughout the Universe.

The Quantum Universe

The Universe could be organised in quantum levels just like the atom. Perhaps we live on a level or plane of energy based on the speed of light. There could be levels of energy based on speeds faster than the speed of light. I call energy beyond light 'super-energy'. Worlds of super-energy, beyond the speed of light, would exist outside the bounds of physical space and time.

Dimensions of Speed

Do we live in a multi-dimensional Universe where different dimensions are separated by the speed of their energy? Is the speed of light the speed of energy on our level? Is this why everything in our world is relative to the speed of light?

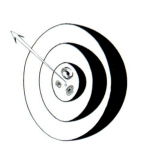

The Cosmic Laws

<u>1. Law of Subsets:</u> Lesser speeds are part of greater speeds so our world of energy would be a sub-set or part of the worlds of super-energy.

<u>2. Law of Simultaneous Existence:</u> Space and time are created by vortices and waves on each plane so there would be no space-time separation between energy and super-energy.

Russian Dolls

Nested Russian dolls show the idea that worlds made of lesser speeds are part of worlds based on greater speeds of energy.

Super Energy

Worlds of super-energy could be all around but unperceived because their velocity of energy would 'relatively' too fast for us.

Angels and Fairies!

Super energy beings all round us, perhaps, in their own super-space and time with us unaware of them because – according to the Theory of Relativity - their 'velocity of light' is too fast? Were we not told there are angels all round us but they move too fast to be seen and there are fairies in the garden that we are too slow to perceive?

Matchbox Mentality

Imagine match people who were aware only of their matchbox. Unaware of the kitchen beyond they deny its existence. In the kitchen there is a cook who could crush them underfoot but chooses instead to leave patterns on the box inviting them to expand their horizons. Sadly the stupid matchbox people dismiss these as a hoax!

The Hoaxer

"How do I make crop circles?"

"Easy, The Skeptics drop me in a field of corn and I flap my arms like so!"

Crop Formations

Huge crop formations of imaginative and complex design appear every season. They appear suddenly and no one is ever caught trespassing! Maybe crop formations reveal intelligence in worlds of which ours is but a part. Whoever creates the formations would obviously be aware of our world even if we were not directly aware of them. The size, diversity and complexity of the patterns suggest the intelligence creating them is smarter than Doug and Dave! The speed and silence with which the patterns appear also suggest a highly advanced technology.

Another Quantum Leap

Just as an electron can take a quantum leap in the atom, maybe bodies of physical energy could take a quantum leap into worlds of super-energy.

Ascension

Bodies should be able to ascend the series of quantum realities from energy into super-energy if the speed of energy in every sub-atomic vortex were to accelerate beyond the speed of light. Should this occur the body would vanish from physical space-time and appear in a world of super-energy without any change in atomic structure or frequency of vibration. (If ascension were an increase in frequency of vibration you could ascend bunny in the microwave!)

The Philadelphia Experiment

During World War II the US Navy conducted an experiment in Philadelphia harbour to see if they could make ships invisible to radar. A vortex turbine was activated in the S.S. Eldridge. The destroyer vanished for fifteen minutes - off the radar screen and out of sight. Did the vortex cause the speed of energy in every atom of the ship and its crew to ascend into hyperspace?

20

Vortex Technology

Inventors including Victor Schauberger, Bruce de Palma, Adam Trombly, Joseph Newman and John Searle claim to have developed anti-gravity, free energy devices from vortex motion. Were they drawing super-energy into our world from hyperspace through a vortex resonance? Did the Germans develop Schauberger's vortex turbines as a secret weapon in World War II?

New Energy and Transport Systems

Vortex technology points to a new future of abundant free energy and levity transport that would render cars and road, rail, ship and existing air systems obsolete, as well as negating the need for oil, gas, coal and nuclear power. For more details visit 'The Cosmic Vortex' on pujapower.com

Free Energy in the News!

March 21, 1986 The Guardian reported that Dr Roger Hastings, chief physicist for the Sperry-Univac Corporation, had tested Joseph Newman's apparatus designed to replace the engine of a Porsche. He found that the production efficiency of Newman's machine was far greater than 100%. Hastings issued an affidavit: "...On September 19th 1985 the motor was operated at 1,000 and 2,000 volts battery input, with output powers of 50 and 100 watts respectively. Input power in these tests were, 7 and 14 watts yielding efficiencies of 700% and 1,400% respectively..." Inventions like this would pose a serious threat to the oil moguls!

UFOS

Vortex technology could explain UFOs, their ability to defy gravity, appear and disappear and generate energy as they travel in and out of space-time, between the worlds of energy and super-energy.

DNA Resonance

The DNA molecule is a spiral. Like a coil in a radio set it could receive frequency information from fields of super-energy. DNA resonance maybe the key to morphic resonance and account for the differentiation of multi-cellular organisms. It could also account for intelligence in the evolutionary changes of DNA.

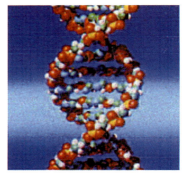

The Field

Snowflakes melted and refrozen reform their original patterns. Photons fired individually through slits still form interference patterns. Many non-biological systems have memory, leaving an imprint of how or where they have been. Perhaps every object, inanimate as well as animate, has a super-energy field, a pattern out of time that leaves a memory of its existence.

Alternative Medicine

Alternative medicine fits well into the new science of consciousness. Acupuncture, reflexology and healing depend upon subtle energy - 'Chi' - that stimulates the body to heal itself. Is it possible that Chi is super-energy acting on the cells through DNA resonance? The super-energy 'morphic field' could be acting as the blueprint for rebuilding the physical body.

Acupuncture

The Chinese discovered Chi energy flow lines were associated with every organ in the body. They called these meridians and found that by stimulating or sedating the flow of Chi in the meridians with heat, pressure or needles they could restore balance in the organs and heal many common ailments.

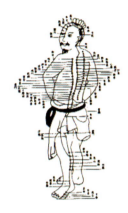

Homeopathy

In 1984 Professor Jacques Benveniste uncovered the basic principle of homeopathy in the laboratory - that a substance can still have a therapeutic effect even when diluted to the extent that not a single molecule of the original substance remains. The homeopathic dilution of a substance increases its super-energy blueprint - something like a photographic negative - so that the homeopathic remedy has the opposite effect to the original chemical. Professor Benveniste discovered water potentised with a drug was just as effective but without side effects!

Homeopathy was invented by Dr Samuel Hahnemann in the 18th Century

23

Planes of Quantum Reality

According to ancient wisdom there are three fundamental planes of reality i.e. Physical, Psychic and Causal (body, mind and spirit). In the new science of consciousness the Universe is considered in three corresponding levels of quantum reality described as the Physical, the Hyperphysical and the Superphysical.

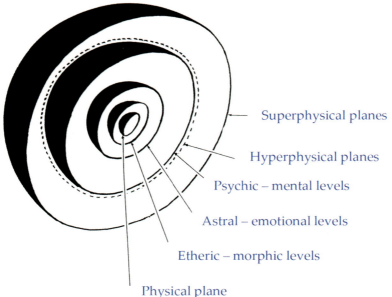

— Superphysical planes

Hyperphysical planes

Psychic – mental levels

Astral – emotional levels

Etheric – morphic levels

Physical plane

The Hyperspherical

The Hyperphysical plane is sub-divided into three levels designated as Psychic (mental), Astral (emotional) and Etheric (morphic).

Making Sense

It is not my intent to prove my ideas by experiment or scientific method. My purpose is to present a speculative picture of the Universe based on scientific concepts for those who believe and experience realities beyond the physical. Many people believe in soul and spirit, higher dimensions and life after death. Many people have spiritual and psychic experiences of other realities. The idea of other quantum realities – levels of energy beyond the speed of light - makes sense of the supernatural, the spiritual and the psychic.

Soul and Spirit

If the Universe has other planes of reality then symmetry suggests the same would apply to us. Maybe we have hyperphysical bodies – a morphic body, an emotional body and a mental body. I use the term soul for the hyperphysical bodies of sensation, feeling and thought. I believe we also have a superphysical body – a vortex of higher thought and transcendental experience through which we are connected to Universal Life Consciousness. This would correspond to Spirit.

Inspiration

With every inspiration of the breath we have the opportunity to connect with Spirit through the still centre of our multi-dimensional 'vortex of being'. As this occurs our higher bodies on each plane, each turn of the spiral, come into alignment and resonate with us. In that timeless moment we are at one with all that is. This happens when we relax and still the mind, focus on the breath and feel into the heart.

25

"There is motion but there are, ultimately, no moving objects; there is activity but there are no actors; there are no dancers, there is only the dance... Like the Vedic seers, the Chinese sages saw the world in terms of flow and change, and thus gave the idea of a cosmic order an essentially dynamic connotation... Shiva, the Cosmic Dancer, is perhaps the most perfect personification of the dynamic universe... The general picture emerging from Hinduism is one of an organic, growing and rhythmical moving cosmos; of a universe in which everything is fluid and ever changing, all static forms being maya, that is, existing only as illusory concepts. This last idea – the impermanence of all forms – is the starting point of Buddhism. The Buddha taught that all compounded things are impermanent', and that all suffering in the world arises from our trying to cling to fixed forms – objects, people or ideas – instead of accepting the world as it moves and changes. The dynamic worldview lies thus at the very root of Buddhism."

Fritjof Capra (The Tao of Physics & Turning Point)

Conclusion

There is nothing to prove, nowhere to go and nothing to achieve. There is only the experience of Life as it is in this moment because now is the only reality we have, so lets live it and enjoy it to the full.